The MOON
Sara Cucini and
Katie Gillespie
EYEDISCOVER

Go to **www.eyediscover.com** and enter this book's unique code.

BOOK CODE

C598822

EYEDISCOVER brings you optic readalongs that support active learning.

Published by AV² by Weigl
350 5th Avenue, 59th Floor New York, NY 10118
Website: www.eyediscover.com

Library of Congress Control Number: 2017930720

ISBN 978-1-4896-5671-1 (hardcover)

Printed in the United States of America
in Brainerd, Minnesota
1 2 3 4 5 6 7 8 9 0 21 20 19 18 17

082017
020317

Editor: Katie Gillespie
Designer: Mandy Christiansen

Weigl acknowledges Getty Images and iStock as the primary image suppliers for this title.

EYEDISCOVER provides enriched content, optimized for tablet use, that supplements and complements this book. EYEDISCOVER books strive to create inspired learning and engage young minds in a total learning experience.

Watch
Video content brings each page to life.

Browse
Thumbnails make navigation simple.

Read
Follow along with text on the screen.

Listen
Hear each page read aloud.

Your EYEDISCOVER Optic Readalongs come alive with...

Audio
Listen to the entire book read aloud.

Video
High resolution videos turn each spread into an optic readalong.

OPTIMIZED FOR

- ✓ TABLETS
- ✓ WHITEBOARDS
- ✓ COMPUTERS
- ✓ AND MUCH MORE!

In this book, you will learn

- how the Moon looks
- what an eclipse is
- how old the Moon is

and much more!

It is night. Look up at the sky! There is the Moon, big and bright.

The Moon moves around Earth. When it moves, its shape looks like it changes.

8

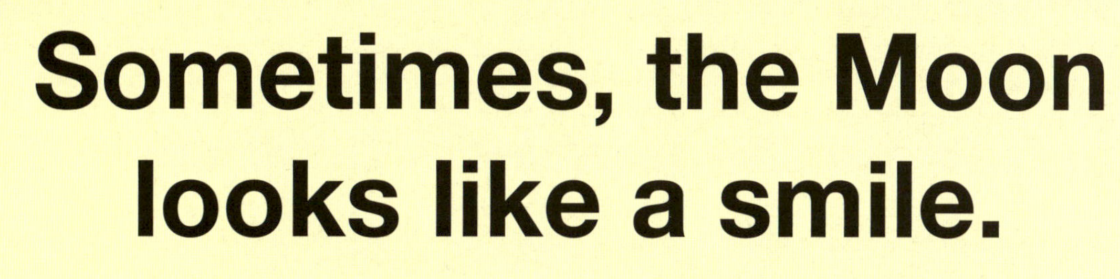

Sometimes, the Moon looks like a smile.

When the Moon's path crosses Earth's shadow, the Moon looks dark. This is called an eclipse.

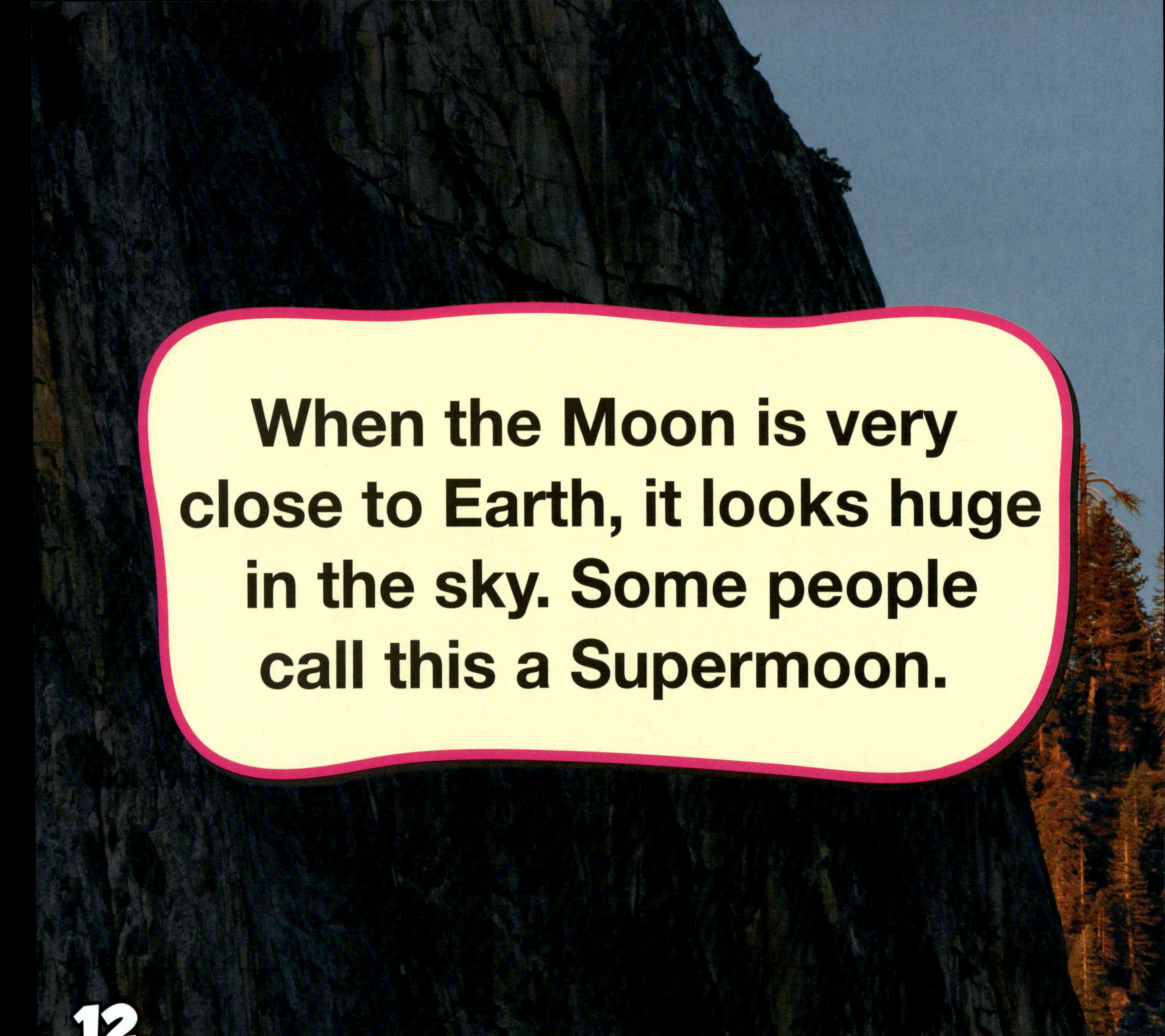

When the Moon is very close to Earth, it looks huge in the sky. Some people call this a Supermoon.

1919

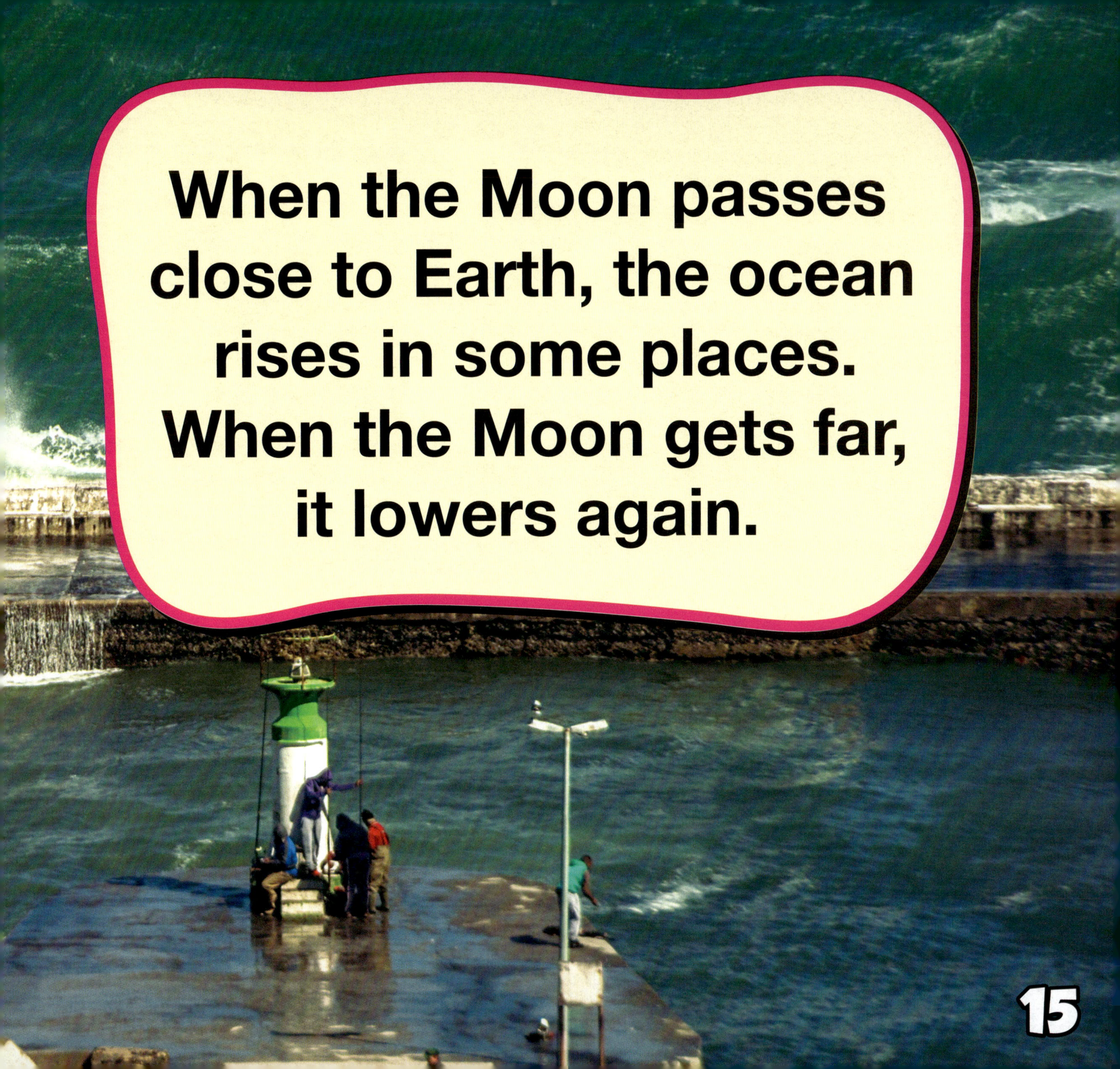

When the Moon passes close to Earth, the ocean rises in some places. When the Moon gets far, it lowers again.

The Moon has many holes and mountains. People have taken pictures and drawn maps of them.

People traveled to the Moon six times to explore it. A spacecraft can reach the Moon in about three days.

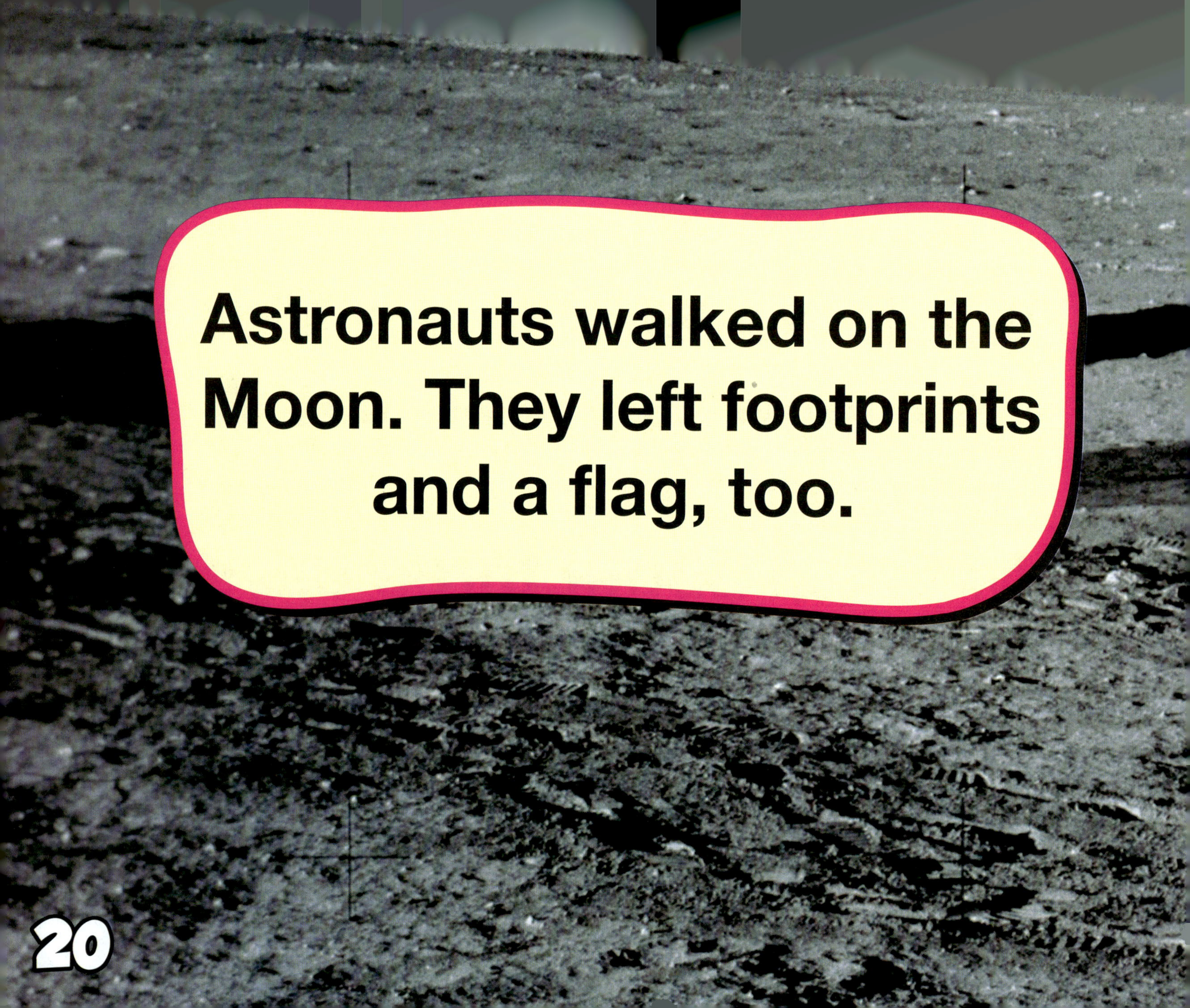

Astronauts walked on the Moon. They left footprints and a flag, too.

The Moon is about
4.5 billion
years old.

We **can only see** about
59 percent
of the Moon **from Earth**.

The Moon is
238,857 miles
away from Earth.
(384,403 kilometers)

On the Moon, you would weigh only **1/6 of your weight on Earth**.

Astronauts left **footprints** on the Moon that will stay there for **10 million years**.

The Moon can be **as hot as 260° Fahrenheit** (127° Celsius) in the **daytime**. It reaches temperatures **as cold as -280°F** (–173°C) **at night**.

KEY WORDS

Research has shown that as much as 65 percent of all written material published in English is made up of 300 words. These 300 words cannot be taught using pictures or learned by sounding them out. They must be recognized by sight. This book contains 51 common sight words to help young readers improve their reading fluency and comprehension. This book also teaches young readers several important content words, such as proper nouns. These words are paired with pictures to aid in learning and improve understanding.

Page	Sight Words First Appearance
4	and, at, big, is, it, look, night, the, there, up
6	around, changes, Earth, its, like, moves, when
9	a, sometimes
10	an, this
12	call, close, in, people, some, to, very
15	again, far, from, gets, places
16	has, have, many, mountains, of, pictures, taken, them
19	about, can, days, three, times
20	left, on, they, too, walked

Page	Content Words First Appearance
4	Moon, sky
6	shape
9	smile
10	eclipse, path, shadow
12	Supermoon
15	ocean
16	holes, maps
19	six, spacecraft
20	astronauts, flag, footprints

Watch
Video content brings each page to life.

Browse
Thumbnails make navigation simple.

Read
Follow along with text on the screen.

Listen
Hear each page read aloud.

Go to www.eyediscover.com and enter this book's unique code.

BOOK CODE

C598822